☞ W9-AJT-076

OXYGEN KEEPS YOU ALIVE

OXYGEN KEEPS YOU ALIVE

BY FRANKLYN M. BRANLEY

ILLUSTRATED BY DON MADDEN

Thomas Y. Crowell Company

New York

LET'S·READ·AND·FIND·OUT SCIENCE BOOKS

Editors: *DR. ROMA GANS*, Professor Emeritus of Childhood Education, Teachers College, Columbia University

DR. FRANKLYN M. BRANLEY, Chairman and Astronomer of The American Museum–Hayden Planetarium

*AVAILABLE IN SPANISH

L.C. Card 73-139093

ISBN 0-690-60702-4
      0-690-60703-2 (Lib. Ed.)

2   3   4   5   6   7   8   9   10

OXYGEN KEEPS YOU ALIVE

When you are awake, you breathe.

When you are asleep, you breathe.

1

Every minute of every day you breathe in air. Without it, you could live only a few minutes. Part of the air that you breathe is oxygen. You cannot see the oxygen, any more than you can see the air. But there is oxygen in every breath you take. For living things, oxygen is the most important part of the air.

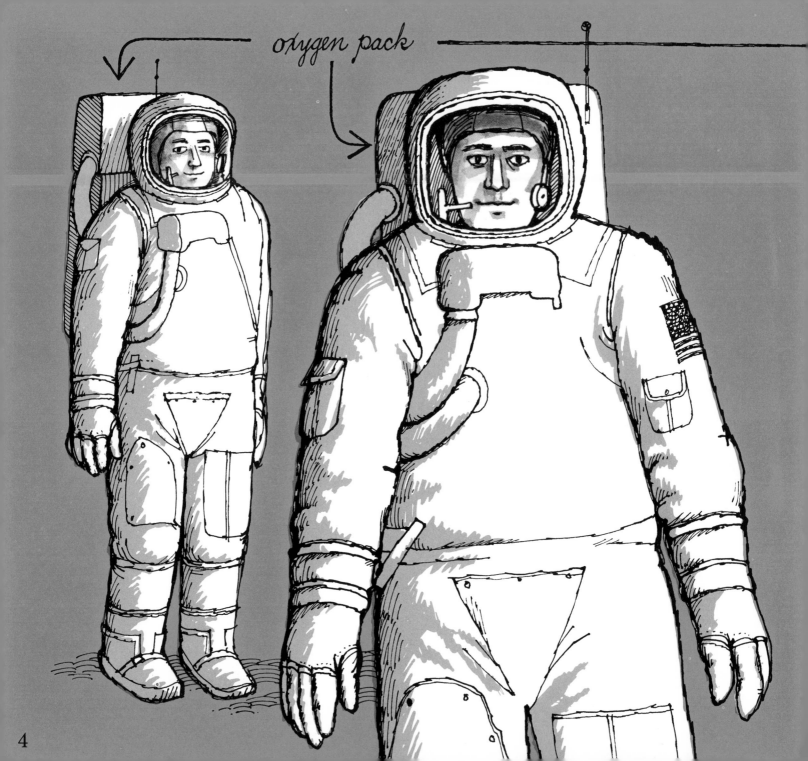

oxygen pack

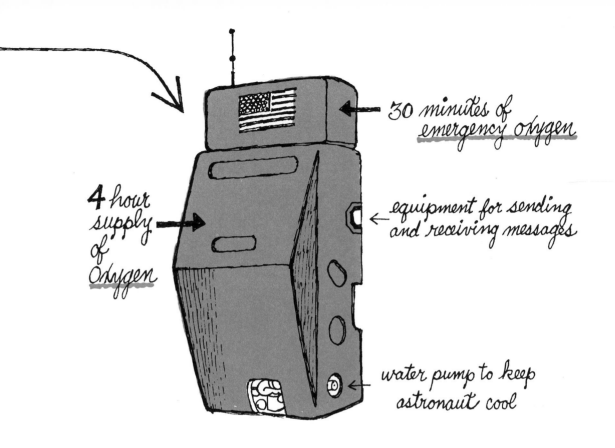

30 minutes of _emergency oxygen_

4 hour supply of _Oxygen_

equipment for sending and receiving messages

water pump to keep astronaut cool

Astronauts in their space suits breathe oxygen. The oxygen keeps them alive. It keeps alive everything that lives.

Dogs and cats, fish and frogs, whales and tigers and fleas, need oxygen. All animals need oxygen. Plants need it. And so do you.

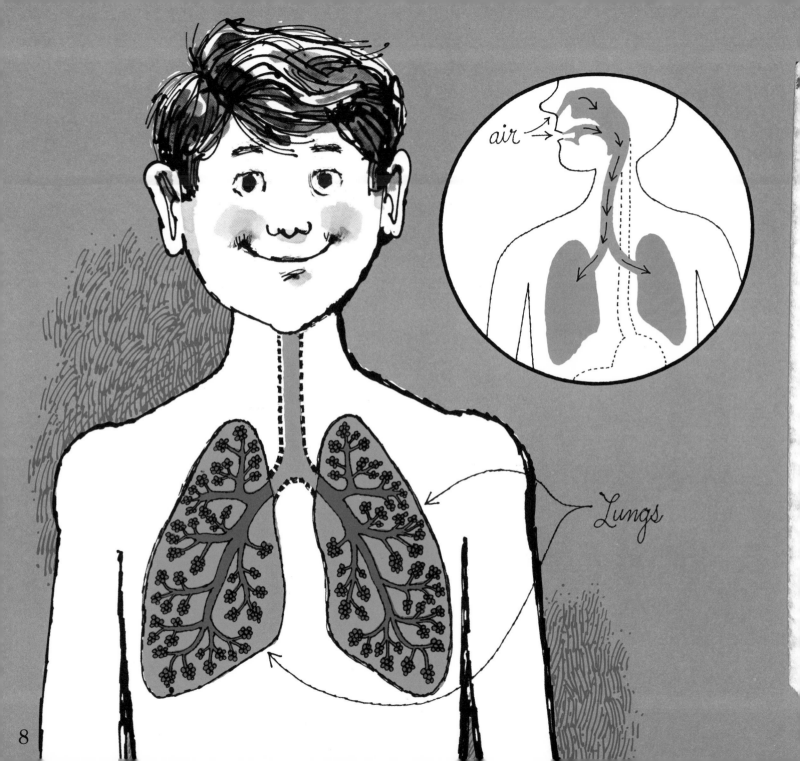

air

Lungs

8

The air you breathe contains many things. It is made up mostly of nitrogen and oxygen. Oxygen is the most important part of the air for you. When you breathe, air goes into your body. Most of the oxygen stays in your body, where it is used. The unused part of the air goes out when you breathe out.

When you breathe in, the oxygen goes through your mouth and nose, and into your lungs. Your lungs are made of tubes that lead to many smaller tubes, which lead to even smaller tubes. They look rather like the branches of a tree. At the ends of the tubes are many small chambers that look like clusters of bubbles. They are surrounded by blood vessels.

Oxygen goes into your blood through the walls of these chambers.

Your blood carries the oxygen to all the cells of your body—in your hands, in your toes, in your eyes. Every cell needs oxygen to stay alive.

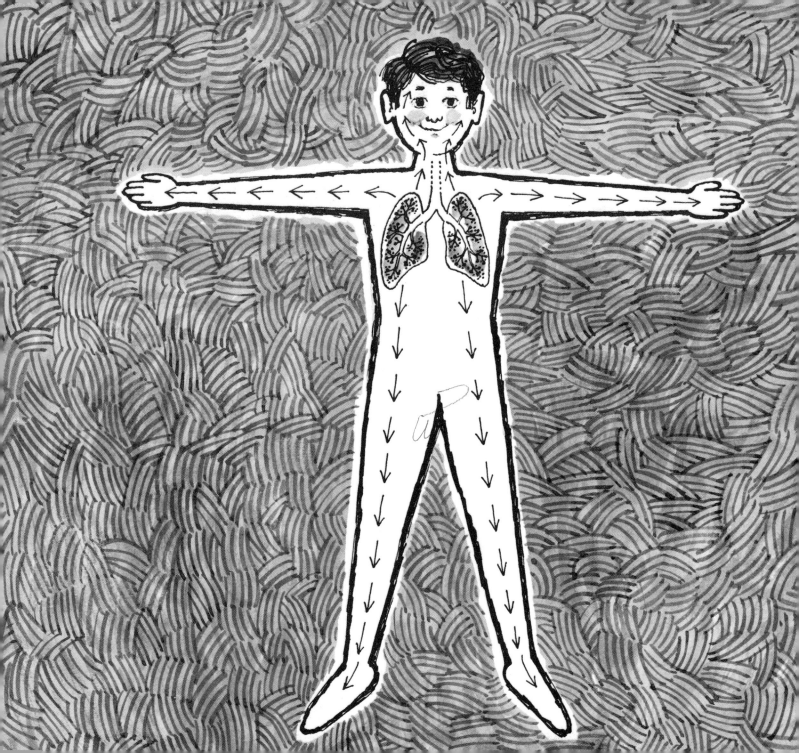

Near the surface of the earth where we live there is
plenty of oxygen in the air. But there are some
places where there is not much oxygen.
Wherever man goes he must have oxygen.
On the tops of high mountains there is not enough
oxygen to keep a man alive. Mountain climbers
must carry tanks of oxygen with them.

When an astronaut walks on the moon in a space suit, he carries a tank of oxygen.

Inside a spaceship astronauts do not need their own oxygen tanks. The whole ship is supplied with oxygen.

Scuba divers who explore under the sea carry tanks of air on their backs. Scuba means *S*elf-*C*ontained *U*nderwater *B*reathing *A*pparatus. Part of the apparatus is a mask that fits over the diver's eyes and nose and keeps the water out. The diver does not breathe through his nose. He breathes through a tube that leads from the tank and into his mouth. Air goes from the tank to the diver. Oxygen in the air keeps the diver alive.

But how about fish? They don't have tanks of air. Fish get the oxygen they need out of the water they swim in. Air with oxygen in it is dissolved in the water.

gill

top view of gills

Fish don't have lungs for breathing. They have gills instead. The gills are filled with tiny blood vessels. Next time you see the gills of a fish, notice how red they are. The mouth and gills move all the time. Water is pumped over the gills day and night, night and day—as long as the fish is alive. Oxygen dissolved in the water goes through the walls of the gills and into the blood of the fish.

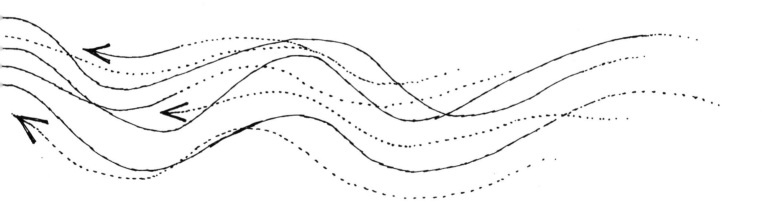

You can do an experiment that shows air dissolved in water. Fill a glass with water. Cover the glass, so air from outside cannot get into the water. After two or three hours, you will see bubbles on the inside of the glass. They are bubbles of air, air that was dissolved in the water. Part of the air in the bubbles is oxygen.

Ask your mother to fill a glass with boiled water. Cover this glass as you did the other one. You won't see any air bubbles in this glass. When you boil water, you drive out the air. That means you drive out the dissolved oxygen too.

A fish cannot live in water that has been boiled, even after the water has been cooled. There is no dissolved oxygen for him to breathe.

All living things need oxygen. Fish and birds need it. So do ants and caterpillars, lions and tigers. All the animals need it. And so do the plants.

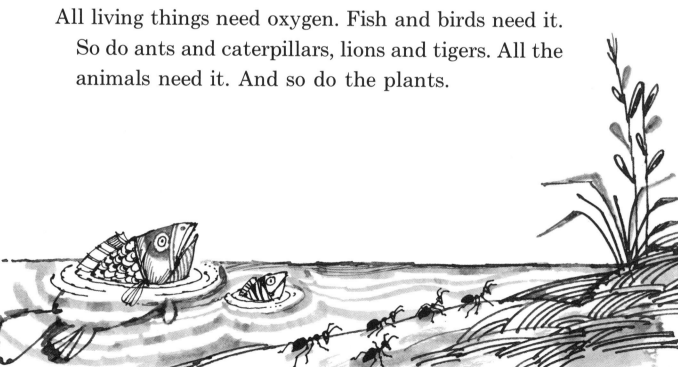

Plants do not breathe with noses and lungs, and they do not have gills. But there are small holes in the underside of a leaf. They look like this under a microscope.

Air, with the oxygen in it, goes into the leaf through these holes. At night plants use some of the oxygen to stay alive, in almost the same way that an animal does.

Plants take in water through their roots. There is oxygen in water. In daytime plants use some of this oxygen to stay alive.

But plants do something else with oxygen that is very important. Green plants use some of the oxygen to make food. They make beans, apples, rye, oats, barley, and other foods that animals eat.

All the food you and I eat comes from plants, or from animals that eat plants. If there were no oxygen, there would be no plants, and if there were no plants, we would have no food.

Man must have oxygen every minute of his life—awake or sleeping. And man must have an oxygen supply wherever he goes—high in the air, deep in the sea, on the moon, on Mars, on Mercury, or on any other planet of our solar system.

Oxygen is important.

ABOUT THE AUTHOR

Dr. Franklyn M. Branley is well known as the author of many excellent science books for young people of all ages. He is also co-editor of the Let's-Read-and-Find-Out Science Books.

Dr. Branley is Chairman and Astronomer of The American Museum–Hayden Planetarium in New York City. He is director of educational services for the Planetarium, where popular courses in astronomy, navigation, and meteorology are given for people of all ages. He is interested in all phases of astronomy and the national space program, and he instructs young people, adults, and teachers in these subjects.

Dr. Branley holds degrees from New York University, Columbia University, and from the New York State University College at New Paltz. He lives with his family at Woodcliff Lake, New Jersey.

ABOUT THE ILLUSTRATOR

Don Madden attended the Philadelphia Museum College of Art on a full scholarship. Following graduation, he became a member of the faculty as an instructor in experimental drawing and design. The recipient of gold and silver medals at the Philadelphia Art Directors' Club exhibitions, Mr. Madden's work has been selected for reproduction in the *New York Art Directors' Annual*, in the international advertising art publication, *Graphis*, and in the *Society of Illustrators Annual*.

Don Madden has always loved animals and the outdoors. He is delighted, therefore, to be living in an old house in upstate New York where he shares the country oxygen with his wife, an artist also, his two children, a golden retriever, 25 assorted chickens and roosters, a Nubian goat, and a black sheep named Josephine.